Make It Move!

Susan Canizares
Betsey Chessen

Scholastic Inc.
New York • Toronto • London • Auckland • Sydney

Acknowledgments
Literacy Specialist: Linda Cornwell
National Science Consultant : David Larwa

Design: Bryce Schimanski
Photo Research: Amla Sanghvi
Endnotes: Mary Hart
Endnote Illustrations: Craig Spearing

Photographs: **Cover:** Jim Erickson/Stock Market; pg. 1: Bob Thomas/Tony Stone Images; pg. 2: Elizabeth Hathon/Stock Market; pg. 3: Bob Daemmrich/Stock Boston; pg. 4: Peter Correz/Tony Stone Images; pg. 5: Gale Zucker/Stock Boston; pg. 6: Bob Thomas/Tony Stone Images; pg. 7:Jim Erickson/Stock Market; pg. 8: Ben Mendlowitz/Stock Market; pg. 9: Felicia Martinez/Photo Edit; pg. 10: Myrleen Ferguson Cate/Photo Edit; pg. 11: Peter Vandermark/Stock Boston; pg. 12: Mike Timo/Tony Stone Images.

No part of this publication may be reproduced in whole or in part, or stored in a retrieval system, or transmitted in any form or by any means, electronic, mechanical, photocopying, recording, or otherwise, without written permission of the publisher. For information regarding permission, write to Scholastic Inc., 555 Broadway, New York, NY 10012.

Library of Congress Cataloging-in-Publication Data
Canizares, Susan 1960-
Make it move!/Susan Canizares, Betsey Chessen.
p.cm. --(Science emergent readers)
Summary: Simple text and photographs present people making things move, from the bouncing of balls to the pedaling of bikes.
ISBN 0-439-08121-1 (pbk.: alk. paper)
1. Force and energy--Juvenile literature. [1. Force and energy.
2. Mechanics.] I. Chessen, Betsey, 1970-. II. Title. III. Series.
QC73.4.C36 1999

531'.6--dc21

98-53313
CIP AC

Copyright © 1999 by Scholastic Inc.
Illustrations copyright © 1999 by Scholastic Inc.
All rights reserved. Published by Scholastic Inc.
Printed in the U.S.A.

7 8 9 10   08   03 02

How can you move it?

Pull it.

Spin it.

Throw it.

Hit it.

Kick it.

Bounce it.

Row it.

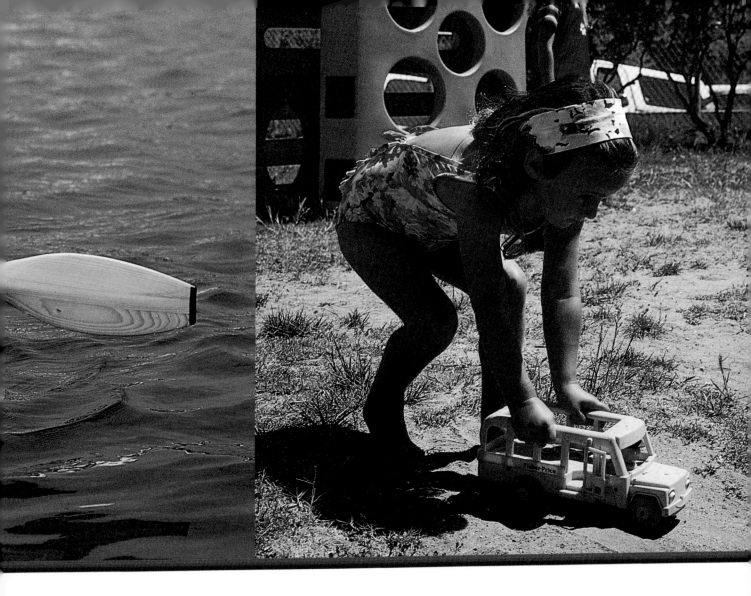

Push it.

Pedal it.

Lift it.

You can make it move!

Make It Move!

How can you move it? Have you ever played baseball or basketball? If so, you were doing something that involves force and work. Things move when something pushes or pulls them. We call something work when there is a transfer of energy from you to something else. These forces can make something start to move, speed up, slow down, change direction, or stop moving. Forces make things move.

Pull it When you pull a wagon, the wheels turn and the wagon moves with you. The wheels on a wagon make it easy to move heavy loads. You can pull a weight that is too heavy to lift off the ground. You can also give your friends a ride!

Spin it When you wiggle your hips, the hoop twirls around your waist. With enough energy and practice, some people can keep the hoop spinning for hours at a time without letting it fall to the ground.

Throw it When you transfer your energy into a ball, you can throw it far! The farthest baseball throw was made by Glen Gorbous in 1957. He threw the ball 445 feet and 10 inches! How far can you throw a baseball?

Hit it It would be hard to break open a piñata using only your fist. But when you transfer your energy into a stick, you can give it a good, hard whack. When you hit a piñata with a stick, be sure to hit it as hard as you can so the toys and candy inside will burst out and fall onto the floor.

Kick it When you transfer your energy by kicking a ball, it goes flying through the air. In a football game, a team scores points when a teammate kicks the ball over the goalpost. A goalpost is 20 feet high. Can you kick a ball that high?

Bounce it When you hit a ball up and down on the floor with your palm, it bounces. During a basketball game, the ballplayers throw the ball to a teammate or bounce the ball down the court. They can't carry or kick it. A player scores points by shooting the ball into the basket, which is 10 feet above the floor. Can you throw a ball that high?

Row it When you row a boat, you push the paddle against the water. This makes the boat move forward. How fast you go depends on how quickly you move the paddle in and out of the water. The fastest speed for paddling a canoe was set by a German team at the 1992 Olympics—12.98 mph.

Push it If you push a toy like this bus lightly, it will move only a short distance. If you push the bus very hard, it will move much farther. The bus goes farther because you are using more force to move it.

Pedal it When you ride your bike, your feet turn the pedals. The pedals turn the chain, which makes the wheels move, and your bike goes forward. Bicycle races are one the most popular sports in the world. The closest race ever was in 1989, when Greg LeMond (U.S.) completed a 2,030-mile course in 23 days.

Lift it You can use you arms to lift things off the ground, but you can also use equipment like a wheelbarrow. When you lift the handles and push forward, the wheelbarrow will move ahead of you. Wheelbarrows let people carry a very heavy load quite a long distance. The ancient Chinese called the wheelbarrow the wooden ox and sometimes equipped it with sails.

You can make it move Sleds are easy to pull over snow and ice. That's because sleds have metal runners instead of wheels. In the far north, where the ground is covered with snow for much of the year, people travel on sleds pulled by dogs. And in Russia, sleighs called troikas are pulled by reindeer.

All people have energy You can take your energy and transfer it to something else. When you transfer your energy into something else, whether it's a ball, a toy, a bat, or a sled, you are doing work and you can make things move!